BIG TRUCKS
DIGGERS IN 3-D

BY MARK BLUM

chronicle books · san francisco

THE PHOTOGRAPHER EXTENDS HIS DEEP APPRECIATION TO THE FOLLOWING COMPANIES

AND INDIVIDUALS FOR THEIR GRACIOUS SUPPORT IN HELPING ME TO

PHOTOGRAPH THESE MARVELOUS MACHINES:

Caterpillar Inc. and Ric McDaniel; the Caterpillar Tinaja Hills Training Center in Tucson, Arizona,

and in particular Joel Russ, Ken Karpoleon, and June Lampson; the foremen and equipment operators

of Pavex Construction Company, including Mike Lax and Don Shoemaker; Shannon Thomas at

Peterson Tractor of San Leandro; Monterey Peninsula Engineering; Moss Landing Marine Laboratories;

Clint Miller Farms of Watsonville; and last but not least, my editor at Chronicle Books, Jennifer Vetter.

INTRODUCTION

You've probably seen CAT® machines on a construction site, or CAT tractors in a field. But did you know there are more than 30 different kinds of CAT machines? And each one of those machines comes in different models and sizes, from big to absolutely enormous! This book will tell you about some of these machines and what they can do. If you look through the glasses built into the cover you'll be able to see awesome CATERPILLAR® machinery in action... and in 3D!

The pictures in this book are called stereoscopic pairs. Each two photos look the same, but are actually different. They were taken from slightly different places, just like our eyes see from two different places. Your brain combines the two pictures into one 3-D image. Some of these pictures were taken with specially built two-lens stereo cameras. Others were made by using two normal cameras mounted side by side. You can make a 3-D image with one camera by taking one picture and then moving the camera over a couple of inches and taking another. Try it!

5230 FRONT SHOVEL

This gigantic machine weighs 702,000 pounds and can dig up 22 cubic yards of dirt with a single pass of its bucket. It's so big that the operator has to climb a series of ladders to get up to the operator's seat!

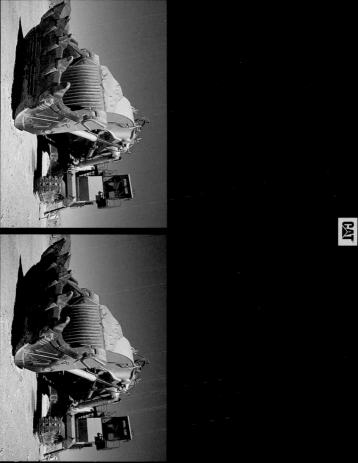

5230 FRONT SHOVEL

The white truck gives a sense of perspective, which is easy to lose with machines this big! The fuel tank of the 5230 Front Shovel holds 1,386 gallons. The bucket can lift 68,000 pounds with ease. That's about the same as two professional football teams plus all the weight they could bench-press (if you could get the quarterbacks to sit nicely next to the defensive linemen!)

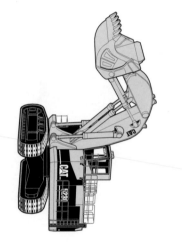

5130B BACKHOE

The 5130B is the next excavator size down from the 5230, and this one is con-figured as a backhoe instead of a front shovel. The hydraulic cylinders run along the top of the boom and stick, the parts that act like an arm. A backhoe bucket scoops back towards the machine and can reach deeper into a hole than a front shovel bucket that scoops out and away.

793C OFF HIGHWAY TRUCK

This enormous, powerful off highway truck can drive over unpaved haul roads for long distances. It can go fast, too, so that workers don't have to spend too much time transporting the truck's load of rocks or dirt from one place to another. This 793C model has a top speed of 33.7 miles per hour (54.3 km/hour) when it's fully loaded, and its load capacity is 240 tons (218 metric tons)! The next size up is the 797. It's the largest mechanical drive truck in the world, and it can go 40 miles per hour while it's carrying 360 tons!

OFF HIGHWAY TRUCKS AND WHEEL LOADERS

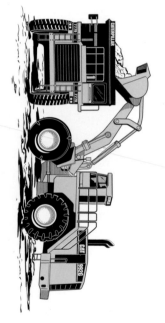

Here in the Arizona desert is a giant parking lot of Caterpillar machines. These off highway trucks and wheel loaders are rugged enough to make their way over rocky or muddy terrain without getting stuck. One of these big off highway trucks could also drive on a regular highway, except that it would take up more than two lanes. Talk about a WIDE LOAD!

992G WHEEL LOADER

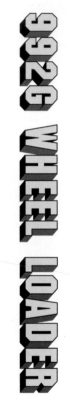
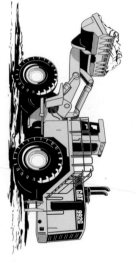

This gigantic front end wheel loader is built for picking up rocks, dirt and debris and loading it into an off highway truck. There are matching sizes of front loaders, excavators, and off highway trucks. A big wheel loader can load a small truck, but you can't use a small wheel loader with a big truck because it has to be able to reach over the side. And it would take many, many bucket loads to fill the truck even if it *could* reach. This 992G Wheel Loader has a bucket range of 15–16 cubic yards, which means that it can load a 777D truck that holds 100 tons in four passes.

854G WHEEL DOZER

Standing on top of one of these gigantic machines in the Arizona desert feels like standing on the top deck of a ship on the ocean. The black painted exhaust pipes even look like periscopes.

950G WHEEL LOADER

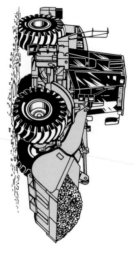

This tractor is helping to rebuild the famous Moss Landing Marine Laboratories, which were destroyed by the 1989 Loma Prieta earthquake in northern California. The sharp leading edge of the blade cuts through rocky dirt and underbrush so the tractor can bulldoze and load dirt and debris. The big wheels with rubber tires help keep the machine body riding high so it doesn't get stuck on mounds of dirt, and they also allow this machine to operate on country roads and city streets from one field or construction site to another.

TRACK-TYPE TRACTOR

This powerful machine also pushes dirt and debris in front of its slightly curved dozer blade. When bulldozing action isn't needed, the blade can be lifted up out of the way while the tractor backs up and moves to other parts of the site. The tracks grip the ground and can travel over the roughest terrain without getting stuck the way wheels might. But these heavy metal tracks would damage a paved street or highway, so a transport truck is required to move the tractor to another site.

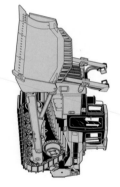

TRACK MACHINES

Track-type tractors, track loaders, front shovels and excavators have metal tracks instead of wheels. Tracks are better than wheels at handling really tough terrain. In fact, they look like the tracks on military tanks, which is no accident. Farming tractors and military tanks were developed around the same time, in the early 1900s, and they used the same ideas and technology.

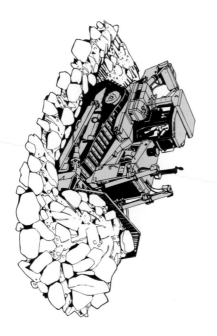

CHALLENGER AGRICULTURAL TRACTOR

In 1982, Caterpillar introduced belted rubber tracks with their Challenger agricultural tractors. Unlike builders, who actually want to pack down the dirt, farmers want just the opposite, because it's hard for seeds to sprout and plants to grow in compacted dirt. Distributing the weight of a tractor over the broad area of the rubber treads helps prevent unwanted soil compaction.

CHALLENGER TRACTOR
WITH ATTACHMENTS

Agricultural tractors can pull plows, fertilizer spreaders, planters or sprinklers behind them. The Challenger above is pulling a large disc plow to turn over the earth in a field of lettuce at a farm in northern California. Can you imagine how long it would take to plow this same field the old-fashioned way, one row at a time, with a horse or a team of oxen pulling a wooden plow?

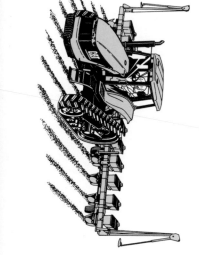

CAT

322 BL EXCAVATOR

The large excavator is a big digger that scoops up dirt and rocks. The bucket has teeth that help cut into hard ground. The boom and stick allows a wide range of motion so that this excavator can reach all the way down to dig into a river-bottom and then stretch all the way up to deposit the material in an off highway truck, for example.

345B DEMOLITION EXCAVATOR

This excavator is removing debris from a demolished building in San Francisco. An excavator can dig with its bucket, break up asphalt or cement with a hydraulic hammer attachment, and lift large blocks of material with a cable attachment. The body of the excavator can swing 360° around, and the tracks can carry it over tough, uneven ground, making this an especially versatile, powerful machine.

301.5 MINI EXCAVATOR

With the auger attachment, this mini excavator becomes a post hole digger. Imagine how much time this machine saves for people building fences! Other possible attachments for the 301.5 are digging buckets, ditch cleaning buckets and hydraulic hammers. It can do everything a big excavator can, but on a miniature scale!

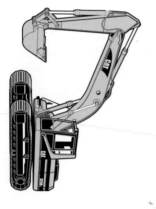

STANDARD EXCAVATOR WITH BACKHOE BUCKET ATTACHMENT

226 SKID STEER LOADER

This is another small machine that is handy for getting into small spaces on construction sites where big machines won't fit. It can do all sorts of things by using an amazing variety of attachments, including utility buckets, dirt buckets, pallet forks, augers, brooms, grapple forks, hammers, landscape rakes, trenchers and compactors.

416C BACKHOE LOADER

This backhoe loader is two machines in one: a front loader at one end and a backhoe at the other. The operator's seat can swivel around to face the steering wheel and the controls for the loader in the front or the backhoe controls in the back.

416C BACKHOE LOADER

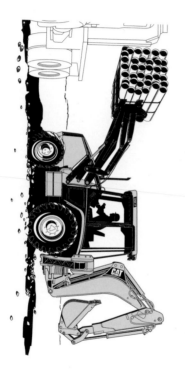

The pair of metal feet at the backhoe end are stabilizers put down to hold the machine steady while the backhoe digs. The hydraulic cylinders push the sta- blizers down until the wheels aren't even touching the ground. The stabilizers can be stored up out of the way when the operator wants to use just the front loader or drive the machine on its four wheels.

24H MOTOR GRADER

When the motor grader is at work, the blade in the middle of the machine is lowered so that it scrapes along the ground, evening out the dirt. The operator angles the blade so that the scraped up dirt is deposited along one side or the other. Motor graders are used for building roads and smoothing out the ground for airport runways, sports playing fields, construction sites, or any place a rough, uneven piece of ground needs to be made smooth and flat. This huge 24H Motor Grader works to clear roads big enough for giant off highway trucks.

816 SOIL COMPACTOR

Capable of dozing, filling and compacting, the soil compactor is useful in road-building and on other construction sites. The tamping-foot wheels work great for traction, gripping the surface without slipping, and compaction, packing the dirt down. Landfill compactors are similar, with sharper tamping-feet on the drums for crunching garbage down so that it takes up less space with fewer air pockets.

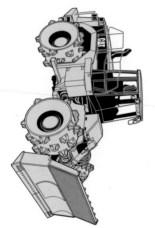

LANDFILL COMPACTOR

CAT

CS563C VIBRATORY DRUM COMPACTOR

Also known as a road roller, this machine puts a smooth, flat surface on graded soil. With its big round smooth drum with heavy weights inside, it compacts just about anything it rolls over. Strong thrumming vibrations allow this machine to pack down soil with more power than a static or non-vibrating roller.

AP1000B ASPHALT PAVER & ASPHALT DRUM COMPACTOR

The paver lays down a layer of asphalt over a prepared surface. The double drum asphalt compactor follows behind the paver, smoothing out the hot new asphalt of roads and parking lots. As the paired drums roll over the fresh new asphalt while it's still hot and soft, they flatten and smooth it to make a good new road surface.

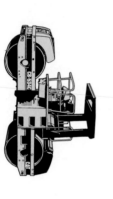

DOUBLE DRUM ASPHALT COMPACTOR

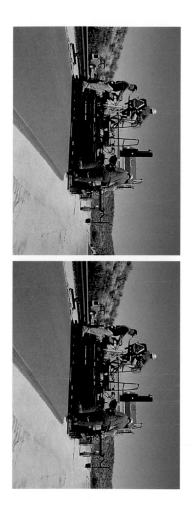

PM565B COLD PLANER

Cold planers scrape off the bumpy, unevenly worn surface of an old road that has potholes, cracks and other problems so that a new surface can be laid down by an asphalt paver. This cold planer can plane a road surface 83 inches wide and up to 12 inches deep.

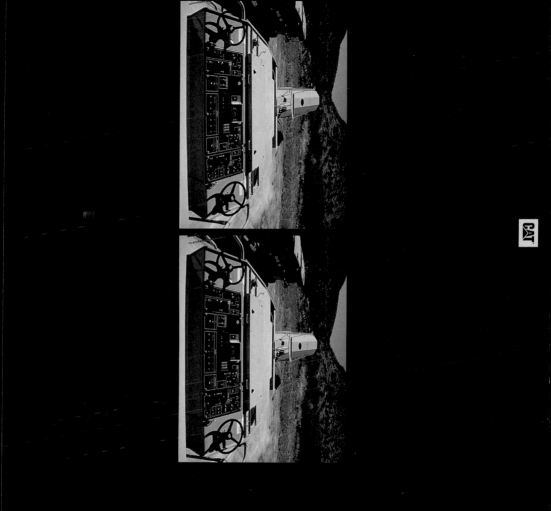

D300E ARTICULATED TRUCK

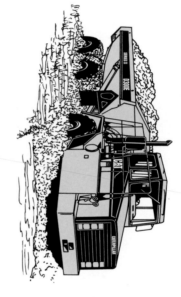

Big trucks often need to make tight turns in small spaces on construction sites, in mines and in logging operations. Articulated trucks have a special flexible joint between the dump body in the back and the engine and cab in the front. This allows it to make a much tighter turn circle than a regular truck. The articulated truck shown here is carrying material scraped up by a cold planer.

WHEEL TRACTOR-SCRAPER

A tractor-scraper can dig and pick up large amounts of dirt in even swathes. So this is the machine for you if you're trying to dig a long shallow trench or clear a path. It can be used to create new roadbeds, and it's especially useful for digging coal or other minerals selectively in an open pit mine. You can even hook two or more of these together.

TH103 TELESCOPIC HANDLER

This machine has an extending arm that stretches out like a telescope to lift material high up in the air, like steel girders up to the third story of a new office building. A steel basket for workers to stand inside can also be attached to the end of the arm so that it acts as a "cherry-picker." This telescopic handler can lift 10,000 pounds as high as 44 feet. That's like lifting five elephants in one load up to the roof of a four-story building.

GLOSSARY

ARTICULATED: able to bend at the middle joint between the truck body and cab.

ASPHALT DRUM COMPACTOR: machine with a heavy, steel drum used to flatten fresh asphalt to make a smooth roadway.

ASPHALT PAVER: large machine with an open hopper that lays down fresh asphalt to pave roads.

BACKHOE: boom and stick with bucket that curves back towards the machine as it digs.

BOOM: the part of the arm closest to the machine body. See "stick".

BUCKET: part at the end of arm that digs and holds material.

CAB: the place where the operator sits and works the controls.

COLD PLANER: large high-powered machine used to mill or grind up pavement and remove tire ruts on the road's surface.

COMPACTION: the process of crunching and crushing material together firmly.

DRUM: a massive steel roller in place of a wheel, used to roll over ground, leaving a smooth surface.

DUMP TRUCK: a motorized machine with a large carrying body that tips up to dump heavy loads of material such as sand and rock.

EXCAVATOR: a track machine with a square body and stick, most often used with a backhoe bucket for digging.

FRONT WHEEL LOADER: wheeled machine with a large front bucket, capable of picking up and loading material into a truck.

GRAPPLER: a metal claw that can grab and drag heavy objects such as tree trunks.

HYDRAULICS: operating by pressurized water or oil.

MOTOR GRADER: a high-powered machine with a long blade that scrapes the ground and pushes aside dirt and gravel to make the ground flat for a new road.

OFF HIGHWAY TRUCK: a giant dump truck that can handle tough terrain.

PLANE: scrape and level the surface.

SCRAPER: machine that digs up a layer of material like coal from the exposed ground surface.

SKID STEER: motorized machine that can make quick turns in tight spaces because its wheels can skid instead of turning in a circle.

STABILIZER: one of a pair of metal feet that hold the machine steady.

STICK: the part of the arm closest to the bucket or other tool attachment.

TAMPING-FEET: these bumps on a compactor's drum are also called sheep's foot pads because they act like the feet of sheep stamping on the ground. The ancient Romans used flocks of sheep to compact dirt roads.

TRACK: steel links or rubber belt put in place of wheels, used to move a machine through tough terrain.

TRACTION: the ability to grip the ground instead of spinning uselessly in place.

TRACTOR: a rugged, powerful machine used to push or pull heavy equipment.

TREAD: the pattern of grooves cut in a rubber tire that help give it better traction.